Bibliographic information published by the German National Library:

The German National Library lists this publication in the National Bibliography; detailed bibliographic data are available on the Internet at http://dnb.dnb.de .

Imprint:

Copyright © 2012 GRIN Verlag
Print and binding: Books on Demand GmbH, Norderstedt Germany
ISBN: 9783668798700

This book at GRIN:

https://www.grin.com/document/441761

Anthony Banyouko Ndah

Understanding the Conflicting Views on Liquefied Natural Gas (LNG) Development Projects and Operations

GRIN Verlag

Understanding the Conflicting Views on Liquefied Natural Gas (LNG) Development Projects and Operations

Anthony Banyouko Ndah (Ph.D.)

Universiti Brunei Darussalam

Contents

Abstract

LNG development is currently amongst the most controversial projects around the world, strongly contested by opponents, generally consisting of environmental activists in communities where LNG operations are planned or on-going, who usually go to great lengths to present to dissuade governments from approving LNG projects, contrary to the views of LNG proponents. Because these conflicting views on the benefits and negative impacts of LNG continue to animate the debate, this paper is intended to examine salient issues of the debate for and against LNG, based on the views of both proponents and critics. The aim is to identify the major sources of the conflicting reactions and perceptions, and propose sustainable solutions for a mutually beneficial and peaceful cohabitation of LNG with the biophysical and social environmental concerns of stakeholder communities.

Two important points have been established:

- That critics of LNG have been vital drivers of innovation in the LNG sector, forcing LNG developers to continuously thrive to design new environmentally friendly technologies.
- That EIA, an invaluable component of all major projects has evolved greatly in the spatial sense, since its introduction in the USA in the 1960s, but its content and methods have changed little over this time. Thus it still dwells strictly on bio-physical and economic considerations, with limited emphasis on social impacts. This is based on the illusion that money can compensate for all other consequences, and especially true of the cases of LNG projects presented here. In most cases the social impacts considered have been limited to such aspects as employment, health, safety, livelihoods, leaving out important cultural, spiritual, relational, emotional or psychological issues; an ominous omission.

This paper concludes that though LNG is a relatively new comer in the global conventional energy sector, it has been growing from strength to strength due to its importance in meeting surging contemporary and future global energy demands, in the face of dwindling oil resources, volatile oil markets, and the high cost of other clean energy sources such as nuclear, solar, and wind energy. It has therefore been recommended that by adopting the guidelines and principles for Social Impact Assessment (SIA), improving techniques of SIA and the inclusion of all major local stakeholders in all stages of LNG projects from planning to implementation (effective stakeholder participation) the rift between LNG development and community resistance could be significantly narrowed.

1. Introduction

Liquefied Natural Gas (LNG), made up essentially of methane gas[1] that has been cooled to approximately -161°C (Sakmar 2010; Foss 2007; USDOE 2009; Al Busaidy 2008) or 261° F [2], at which point it condenses to a liquid for high density transport from places like the Middle East, Russia, and Indonesia where it is produced[3], made its entrance into the international energy scene in 1964 when the first liquefied natural gas (LNG) export was made from the Arzew GL4Z plant (or Camel plant) in Algeria to customers in the UK and France (Al Busaidy 2008). However, the first LNG plant was built in West Virginia, USA, in 1912 and began operation in 1917 (Foss 2007). Initially, most natural gas was consumed in the same region in which it is produced due to the costs and impracticality of transporting natural gas via pipeline over long distances (Sakmar 2010). The natural gas markets have historically been concentrated in the following three distinct regions: North American/Atlantic Basin Region, European Region and the Asia/Pacific Region, and by all estimates, global LNG trade is projected to accelerate over the next 25 years[4]. Over the past decade, LNG has grown to become a significant component of the worldwide energy supply mix and recent reports suggest that global LNG demand is predicted to increase by 10% / year through to 2015[5], especially with the oil and gas-rich Arab states coming into the scene (Al Ghanjah 2011). The fuel is now being developed and shipped or piped around the world in recent times more than ever before, requiring the construction of new facilities such as liquefaction plants and terminals, storage terminals and ports, etc. LNG experienced an unprecedented growth in demand by 22% in 2010, with countries like Qatar producing at its maximum capacity of 77 million tonnes per annum (MTPA) (IGU 2011).

---

[1] Also contains Ethane, Propane, Butane, etc. (see footnote 37 for reference)

[2] Natural Gas Ports and Pipelines : Putting Our Economy , Communities and Quality of Life at Risk available online at www.Nolng.net

[3] Natural Gas Ports and Pipelines : Putting Our Economy , Communities and Quality of Life at Risk available online at www.Nolng.net ; accessed in December 2011

[4] CERA LNG Roundtable 2008

[5] Gulf Times, Aug. 19, 2008

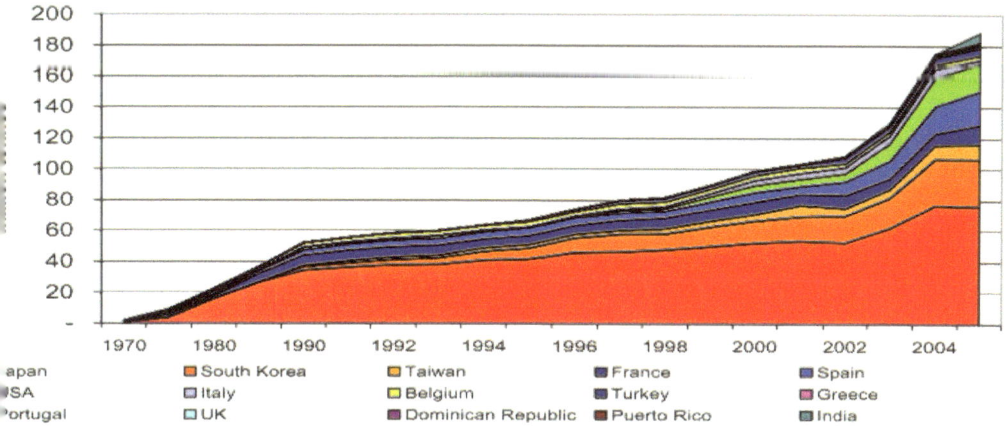

Figure 1. Worldwide Growth in LNG Demand (1970 – 2004)
Source: Cedigaz (1970-1992); Cedigaz, BP Statistical Review of World Energy 2006; USDOE (2008)

The advantage of LNG is that it occupies approximately 1/600th of the volume of natural gas thereby creating more options for shipment and storage (Oman LNG LLC 2010 1, 2 & 3; Sakmar 2010; Al Busaidy, 2008). Figure 1 reveals that since 1970s, Japan has dominated the market demand for imported energy sources as well as LNG, and thus the major client of the LNG business for 30 years; but the size of the market and the number of importers is growing steadily (USDOE 2008). However, the LNG market was poised to experience a down-turn situation in 2011 because major importers like the USA turned from LNG to other domestic unconventional energy sources, as well as diminishing demand from the EU due to sluggish economic growth (IGU 2011). This downturn was however never experienced because the tragic earthquake which hit Japan in March 2011 became a "blessing" for the LNG industry; Japanese municipalities/local authorities temporarily shutdown all the 54 nuclear powered plants which led to the country having to rely on LNG to fill its energy vacuum, resulting in a 12% (about 79 MT) increase in Japan's demand for LNG, and a further global increase in demand of about 8% in 2011 (a new high of 241.5 MT.), with 25 countries actively involved (IGU 2011).

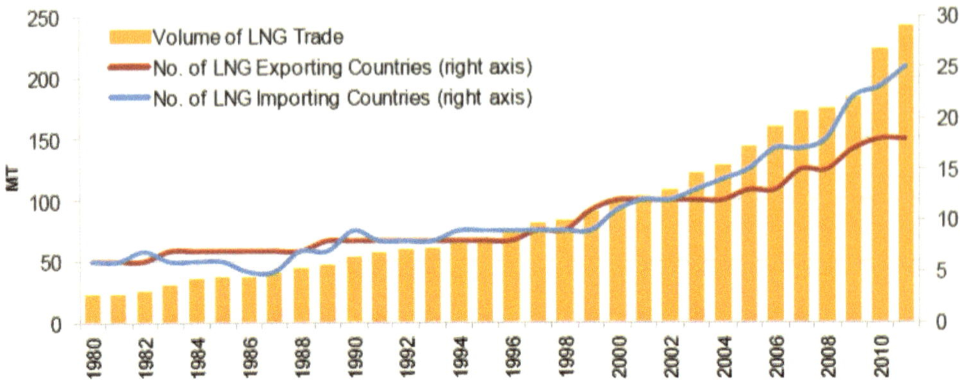

Figure 2. LNG Trade volumes 1980 - 2011
Source: Cedigaz; USDOE; IGU (2011)

On the supply side, global LNG production has been categorized into two broad regional production and supply basins, namely: the Asia-Pacific Basin and the Atlantic Basin (USDOE 2008). In the Asia-Pacific Basin by 2006, Indonesia was the largest LNG producer and exporter, accounting for about 21 percent of the world's total LNG exports, with the majority of its LNG being imported by Japan, with smaller volumes going to Taiwan and South Korea. Also within this basin, Malaysia, the world's third-largest LNG exporter, ships primarily to Japan with smaller volumes to Taiwan and South Korea and about 90 percent of Brunei Darussalam's output goes to Japan (Figure 3).

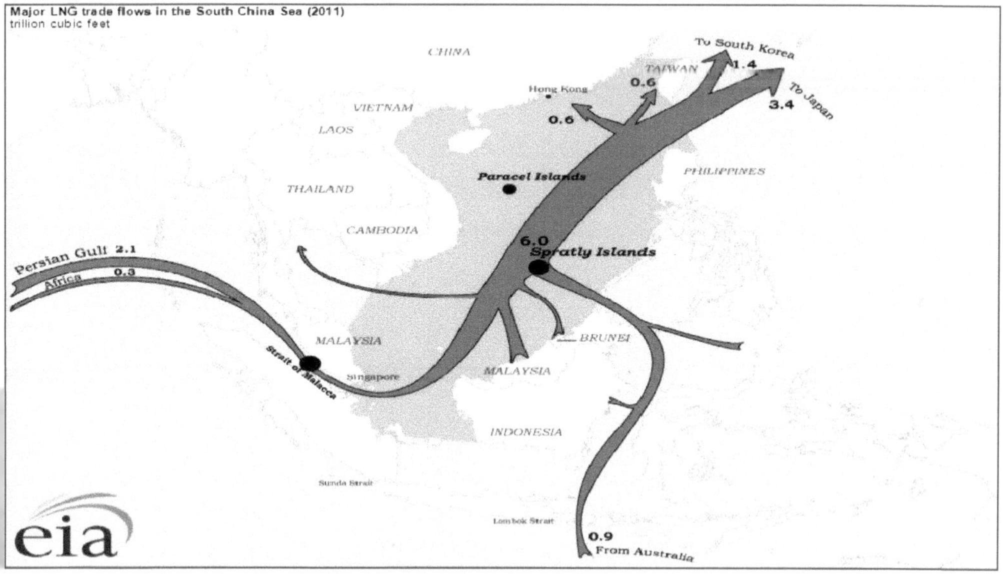

Figure 3. Major LNG trade flows across the South China Sea (2011)

Source: U.S Energy Information Administration (EIA) URL: https://www.eia.gov/maps/

Meanwhile, in the Atlantic basin, Qatar emerged at the top of the global LNG supply chart, slightly exceeding Indonesia. Algeria was the world's third largest LNG exporter serving mainly Europe (France, Belgium, Spain, and Turkey) and the United States via Sonatrach's four liquefaction complexes (USDOE 2008). Russia became the newest Asia/Pacific Basin exporter, with its first LNG plant on Sakhalin Island off the country's east coast, which was inaugurated on 18 February 2009, and had it first cargo loaded onto the LNG carrier *Grand Aniva* at the end of March 2009 (USDOE, 2008; Mosolova and Dyomkin 2009). No new country had joined the club of LNG exporters in 2011 since Peru became the 18th LNG exporter in 2010.

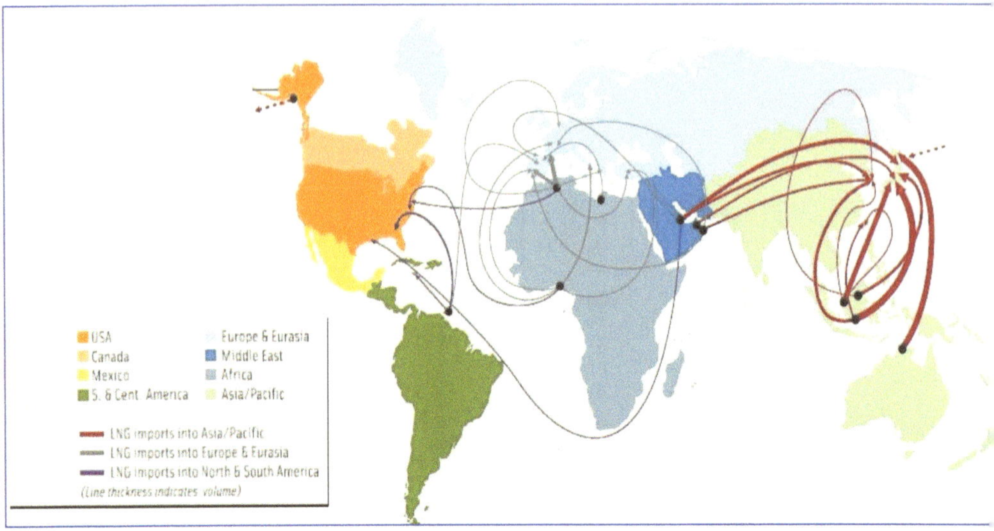

Figure 4. LNG Trading by the geographic region (basins) in which it takes place *(Line thickness indicates volume of trade and arrow heads indicates destination and tails, the source)* *(Source: US Department of Energy – DOE, based on* Energy Information Administration and British Petroleum, URL: https://energy.gov/sites/prod/files/2013/04/f0/LNG_primerupd.pdf)

Figure 4 reveals that the largest volume of trade in LNG takes place in Asia Pacific Basin (Both imports and exports), with S.E. Asia being the largest producer, seconded by the Middle East, while N.E. Asia (predominantly Japan), emerges as the major importer. The Middle East and Africa mainly supply the Atlantic Basin while North America, the less significant supplier but major importer - the USA -supplies mainly Japan from its LNG liquefaction plant in Alaska.

By the end of 2011, the configuration of the global LNG had changed drastically. 18 countries were exporting their gas resources as LNG while five countries, namely Belgium, Brazil, Mexico, Spain and the United States, were re-exporting LNG previously imported from another source. Moreover, Qatar had moved to become the undisputed leader in LNG production, supplying 75.5 MT of LNG to the market in 2011 – nearly one third (31%) of global supply and Malaysia also overtook Indonesia as the second largest LNG exporter in 2011(Figure 4). Together with Australia, these three Pacific Basin exporters accounted for about 27% of the world's LNG supply (IGU 2011). Algeria was relegated to 7th, losing 3 places to Australia, Nigeria and Trinidad and Tobago respectively.

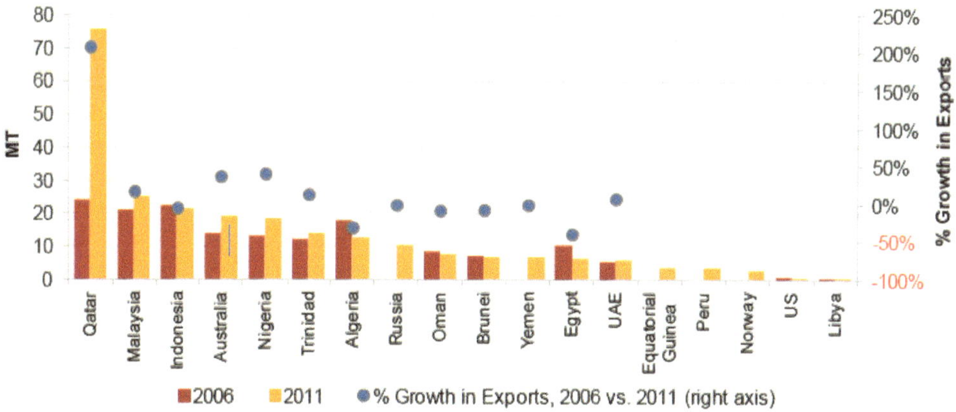

Figure 4. Total production capacity and export of the 18 Global LNG producers 2006 - 2011
Source: US Energy Information Administration (USEIA) and British Petroleum (BP)

LNG development (production and supply) has therefore experienced an exponential growth globally due to improved liquefaction techniques and handling convenience and as a result of these advantages, nations are increasingly viewing LNG as an important fossil fuel for a sustainable future. Considered a clean-burning fuel, many policy leaders and LNG operators have suggested that LNG can play an important role as the world struggles to meet growing energy demands using more environmentally sustainable fuels. However, critics claim that the safety and environmental impacts, including life-cycle emissions of LNG, may nullify any clean-burning benefit LNG might otherwise provide (Sakmar 2010), while communities at large have generally expressed more concerns about environmental and human health and safety issues associated with LNG processes. Based on available literature therefore, this paper attempts to: effectively analyse the conflict in order to enhance understanding on the positions of both camps thereby proposing options that may narrow the rift.

2. Situating the Debate on LNG Development

2.1. Benefits of LNG based on the views of proponents

Proponents of LNG have enumerated ambitious positive environmental, economic and financial benefits to be derived from LNG development, and in their views, LNG is no doubt the fuel that will reduce global over-reliance on the so-called dirty fossil fuels (Oman LNG LLC, 2010 (1, 2, 3); Peru LNG Project (n.d); Calypso LNG project, 2007; etc.). It is generally noted that LNG is clean, odourless, colourless, non-combustible, non-corrosive and non-toxic (Peru LNG project EIA; Kaganawa, 2007; Oman LNG LLC 2008; 2010). Therefore, LNG will not pollute land or water resources, and if LNG is released on water, it evaporates with no residual trace, as well as will not cause an explosion in case of a tank rupture because it is stored at ambient pressure (Peru LNG project EIA, n.d.). This however, can be viewed as an attempt by LNG operators to mask the real danger because in case of a leak, this highly volatile gas might be diluted with air, the natural gas/air mixture would become potentially explosive when it comes in contact with any source of ignition[6].

A major break-through of LNG is that liquefaction enables natural gas that would otherwise be "stranded" to reach major markets, permitting Developing Countries with plentiful natural gas resources and interested in monetizing natural gas to export it as LNG, while developed nations with little or no domestic natural gas rely on such imports (Al Busaidy 2008*)*. According to US Department of Energy (2008), this is an important option in further strengthening globalization through a win-win cooperation.

In the USA for example, (USDOE 2008), the full benefits of LNG in ensuring energy sustainability point to the vital role natural gas plays in energy supply and in achieving the nations' economic and environmental goals especially through increasing imports of LNG to ensure that American consumers have adequate supplies of natural gas in the future[7].
Several reasons advanced for increasing reliance of the U.S. on LNG include:
• Utilities realize advantages by using natural gas-fired generators to create electricity (lower capital costs, higher fuel efficiency, shorter construction lead times, and lower emissions).
• The residential sector benefits from the higher fuel efficiency and lower emissions of gas appliances.
• The industrial sector relies on natural gas as a fuel for manufacturing many of the products we rely on today, including pulp and paper, metals (for computers, automobiles, and telecommunications), chemicals, fertilizers, fabrics, pharmaceuticals, and plastics.
• The transportation sector is beginning to see natural gas as a clean and readily available alternative to other fossil fuels[8].

In another example, economic and financial gains of the Peru LNG project was estimated at roughly US$800 million of hard export currency revenues annually, increasing the country's total exports by an estimated 1.5%. other benefits include: a contribution to GDP growth to make Peru a net hydrocarbons exporter in the medium term; contribution of an estimated US$200 million per year on average in incremental royalties and US$150 million per year in additional income taxes over the first 20 years; yield of US$4.8 billion in net present value terms

[6] Retrieved online at http://www.beachapedia.org/LNG_%28Liquified_Natural_Gas%29 accessed on 08/31/12
[7] U.S.Department of Energy (DOE) in collaboration with the National Association of Regulatory Utility Commissioners (NARUC), in a report titled: Liquefied Natural Gas : Understanding the basics
[8] EIA, Annual Energy Outlook 2005

in cumulative economic benefits and large amounts of foreign direct investment to Peru as well as positively impact the energy matrix in the Latin American region, with gas most likely being shipped to Mexico, thus reducing Mexican dependence on U.S. natural gas[9]. These are just a few examples of the potential and observed benefits of LNG development. However, whether these exciting results and figures are realistic and/or have been achieved in areas where LNG development has taken place or is on-going, remains to be verified.

Generally, in all such ongoing projects, proponents always tend to dwell more on the benefits of the project (financial and economic), with little or limited mention of the negative impacts. Moreover, in the EIA statements of most projects, even less is mentioned about the social impacts of the project on the social environments and livelihoods of the impacted communities. As a result, voices against such projects and critics of LNG have arisen around the World and appear to be equally as vocal and convincing in their arguments as the proponents, focussing on most of the aspects generally missed in the EIAs and/or operational impacts of the said projects, ranging from environmental, social, and cultural issues. Some of the critical views are discussed in the following section.

2.2. Giving voice to the critics of LNG Development

Despite the rapid development of LNG and its claimed benefits and opportunities as put forth by proponents and operators, there is an equally major draw-backs and numerous challenges of natural gas exploitation and trade. Consequently, a diverse group of actors or stake holders, including local communities concerned, environmentalists, scholars, have also emerged who are totally against LNG, and go to great lengths to discredit the sector and call for the complete abandonment of LNG plans or projects.

The construction of pipelines either underground or underwater – a highly expensive and risky undertaking and which ties buyer to seller pretty much irrevocably (Al Busaidy, 2008). This means that, not only does the construction of pipelines come along with great financial and environmental costs, but also, once a pipeline is constructed linking two countries, for the purpose of production and sharing of LNG, such states will still be linked by the pipelines even during conditions of strained political relations. Threats posed by LNG operations range from less severe, severe and very severe, based on different stages of operation and on diverse installations or facilities which according to the IFC-WBG (2007)[10], include LNG base load

[9] Environmental and Social Strategy: Peru LNG Project http://www.perulng.com/

[10] The Environmental, Health, and Safety (EHS) Guidelines are technical reference documents with general and industry specific examples of Good International Industry Practice (GIIP) prepared by International Finance Cooperation (IFC), World Bank Group (WBG) (2007).

liquefaction plants, transport by sea, and re-gasification and peak shaving terminals, harbours, jetties and general coastal facilities. Based on the different structures and processes involved, impacts of LNG may be direct and/or indirect (Reef, B.[11]; Environmental Impact Statement of PNG LNG Project).

Critics of LNG have also based their arguments on the narrative of a significant impact on global warming because the energy used in liquefying, shipping and re-gasifying LNG has a lifecycle green-house gas impact that is 30% or more than domestic natural gas[12]. In the September 2007 edition of the journal Environmental Science and Technology, Carnegie Mellon researchers show that liquefied natural gas (LNG) imported from foreign countries and used for electricity generation could have up to 35% higher lifecycle greenhouse gas emissions than coal used in advanced power plant technologies[13]. Moreover, methane is 20 times more powerful a greenhouse gas than carbon dioxide with a more substantial radiative forcing than CO_2, as well as its potential to increase tropospheric ozone levels, a principal ingredient in "smog," which is harmful to human health and reduces agricultural productivity[14].Foreign LNG has also been found to have many contaminants that domestic natural gas does not and therefore produces more air pollution whether it is being burned by industry, power plants or in household heaters[15,16].

In areas where LNG importation is planned, it is also constantly feared that it would increase dependency on foreign fossil fuels and undermine efforts to address global warming, promote conservation and renewable alternatives[17] [18]. Based on findings, implementation of LNG development plans may result in incremental contributions to cumulative impacts in combination with other past, present, and reasonably foreseeable future marine projects[19].

A summary of potential and/or actual environmental issues and project-specific risks and impacts associated with LNG facilities in coastal environments, which are normally considered as part of a comprehensive assessment and management program are described below[20].

[11] **n.d.** means 'no date', indicative of undated sources or references used in this review

[12] online at http://timrileylaw.com/LNG_FOSSILFUEL.htm from an article titled: Liquefied Natural Gas is Just Another Dirty Fossil Fuel. Accessed on 08/31/2012

[13] online at http://timrileylaw.com/LNG_FOSSILFUEL.htm from an article titled: Liquefied Natural Gas is Just Another Dirty Fossil Fuel. Accessed on 08/31/2012

[14] online at http://timrileylaw.com/LNG_FOSSILFUEL.htm from an article titled: Liquefied Natural Gas is Just Another Dirty Fossil Fuel. Accessed on 08/31/2012

[15] Natural Gas Ports and Pipelines : Putting Our Economy , Communities and Quality of Life at Risk available online at www.Nolng.net

[16] Retrieved online at http://www.beachapedia.org/LNG_%28Liquified_Natural_Gas%29 accessed on 08/31/12

[17] Liquefied Natural Gas : The Wrong Choice for the West Coast. *Energy*, *94104*(415)(2004)

[18] online at http://timrileylaw.com/LNG_FOSSILFUEL.htm from an article titled: Liquefied Natural Gas is Just Another Dirty Fossil Fuel. Accessed on 08/31/2012

[19] Calypso LNG Deepwater Port Project, DEIS ES-1, November 2007November 2007, available online at: *http://www.suezenergyna.com/ourcompanies/documents/CalypsoDEISExec.Summary.pdf*

[20] Environmental, Health, and Safety Guidelines, LNG - LIQUEFIED NATURAL GAS FACILITIES, prepared by International Finance Commission (IFC) and the World Bank Group (WBG)

Ballast water discharges and LNG Vessel cooling: Specific studies reveal that large ships, constructed to transport liquefied natural gas (LNG), need to take on ballast water to stabilize the ship during offloading of the LNG, and a 200,000-m3 capacity LNG carrier would withdraw approximately 19.8 million gallons of water over a 10-hour period at an intake rate of 2 million gallons per hour (FERC 2005). Increasing the temperature of the marine environment artificially[21] by up to 3° C in some cases, from the cooling of LNG tankers is an important source of marine pollution[22]. The effects of these dual processes can be far reaching for the marine environment; significantly altering the equilibrium of the area impacted resulting in ecological disturbances, the direct loss of marine life from areas of water withdrawal or warming, and the introduction of alien species in areas of water discharge, where the LNG is off-loaded.

Concerns over deleterious ecology impacts of LNG: some unavoidable processes of LNG development including: dredging[23], "once through cooling" process of coastal power plants and Seawater Desalination facilities[24], ballast water etc, have been said to entrain eggs and larvae in the ballast tanks of LNG ships at a rate of tens of thousands of gallons per minute[25], change the hydrological regimes, salinities, and flow patterns[26] in coastal areas, impeding of the migration of anadromous fishes and affect the growth and reproduction of filter feeding organisms[27], all having severe adverse consequences for marine ecological health. In the United States, an offshore LNG degasification facility using a closed-loop system proposed near Gloucester, MA, was opposed due to estimated annual mortality of eggs and larvae of Atlantic mackerel, pollock, yellowtail flounder, and Atlantic cod of 8.5 million, 7.8 million, 411,000, and 569,000, respectively, from vessel ballast and cooling water (USCG 2006)[28]; and Calypso LNG project (FERC 2005); estimated catastrophic impingement of large numbers of fish eggs and larvae was also a concern brought about by LNG plans off the coast of Oregon and southeast Florida respectively[29]; in Australia, LNG project in Kimberley coastal waters was opposed on grounds that its impacts outweigh benefits, with far reaching and very damaging effects on the natural

[21] The United Nations and the European Union *Water Framework Directive*, at §2(33) defines pollution as:"... the direct or indirect introduction, as a result of human activity, of substances or heat into the water or land which may be harmful to human health or the quality of aquatic ecosystems or terrestrial ecosystems directly depending on aquatic ecosystems, which result in damage to material property, or which impair or interfere with amenities and other legitimate uses of the environment."

[22] Extract from an article titled Liquefied Natural Gas Ports and Pipelines:Putting Our Economy, Communities and Quality of Life at Risk. Oregon citizens against the pipelines standing up for rural communities; available online at www.Nolng.net ; accessed in December 2011

[23] PHYSICAL EFFECTS—WATER INTAKE AND DISCHARGE FACILITIES: CHAPTER EIGHT, Accessed online at: http://www.nefsc.noaa.gov/publications/tm/tm209/pdfs/ch8.pdf ; accessed in December 2011

[24] Retrieved online at http://www.beachapedia.org/LNG_%28Liquified_Natural_Gas%29 accessed on 08/31/12

[25] PHYSICAL EFFECTS—WATER INTAKE AND DISCHARGE FACILITIES: CHAPTER EIGHT, Accessed online at: http://www.nefsc.noaa.gov/publications/tm/tm209/pdfs/ch8.pdf

[26] PHYSICAL EFFECTS—WATER INTAKE AND DISCHARGE FACILITIES: CHAPTER EIGHT, Accessed online at: http://www.nefsc.noaa.gov/publications/tm/tm209/pdfs/ch8.pdf

[27] PHYSICAL EFFECTS—WATER INTAKE AND DISCHARGE FACILITIES: CHAPTER EIGHT, Accessed online at: http://www.nefsc.noaa.gov/publications/tm/tm209/pdfs/ch8.pdf

[28] Extract from an article titled Liquefied Natural Gas Ports and Pipelines:Putting Our Economy, Communities and Quality of Life at Risk. Oregon citizens against the pipelines standing up for rural communities; available online at www.Nolng.net

[29] *http://www.suezenergyna.com/ourcompanies/documents/CalypsoDEISExec.Summary.pdf Calypso LNG Deepwater Port Project, DEIS* ES-1 *November 2007; accesses in January 2012*

and cultural values of Kimberley, which is a world class marine wonderland[30] (Reef, B. (n.d.)); the Australia Pacific LNG project in the Curtis Coast region, was also opposed because it affected conservation areas including wetlands, seagrass beds, turtle nesting beaches, shore bird roosting areas, coral cays, and planar reefs as well as the Great Barrier Reef World Heritage Area (GBRWHA)[31].

Intensification of coastal use conflicts has also been cited as a basis for opposing proposed LNG projects, because LNG terminals are major industrial facilities, like other coastal development projects, and may therefore incompatible with existing coastal environmental objectives and other coastal-dependent uses (such as fisheries, tourism)[32]. In Sakhalin Island, Russia, extraction of oil and gas by Shell and Exxon Mobil in a pristine marine environment has been criticized because this area is home to the critically endangered Western Pacific Gray Whale, as well as causing massive fish die-offs, two significant oil spills, and contamination of the local water supply; the potential for "wet" unprocessed gas to burn hotter and cause power plants to unintentionally spew more pollutants to the atmosphere, if not properly calibrated[33].

Human safety concerns have also been raised as a justification for the opposition to LNG projects, such as the potential for explosions; three LNG import terminal projects in the California coast were termed a giant step in the wrong direction for the West Coast[34]; consequently, the concern over human safety and the risk of explosions was a main reason one of the proposed LNG terminal in Long Beach, California was rejected by the Long Beach City Council[35]. In Oregon and Florida, public safety concerns and environmental impacts were also raised in protest to the Oregon LNG project[36]; The LNG project in Camisea gas fields in Peru was also opposed based on potential impacts on the environment and livelihoods of riverine indigenous communities[37]. In other to accommodate and limit opposition to LNG projects, thereby maintaining a good reputation and enhancing societal acceptability, significant innovations have occurred in LNG operations around the World.

---

[30] Reef, B. (n.d.). 10 Reasons why the Browse LNG development should not go on the Kimberley coast: www.wilderness.org.au/kimberley accessed in December 2011
[31] Coastal Environment Technical Report - LNG Facility and Pipeline, Australia Pacific LNG Project EIS; Volume 5: Attachment 27
[32] http://www.beachapedia.org/LNG_%28Liquified_Natural_Gas%29 ; accessed in January 2012
[33] online at http://timrileylaw.com/LNG_FOSSILFUEL.htm from an article titled: Liquefied Natural Gas is Just Another Dirty Fossil Fuel. Accessed on 08/31/2012
[34] Liquefied Natural Gas : The Wrong Choice for the West Coast. *Energy*, *94104*(415)(2004)
[35] http://www.beachapedia.org/LNG_%28Liquified_Natural_Gas%29 ; accessed in January 2012
[36] Natural Gas Ports and Pipelines : Putting Our Economy , Communities and Quality of Life at Risk available online at www.Nolng.net
[37] Liquefied Natural Gas : The Wrong Choice for the West Coast. *Energy*, *94104*(415)(2004)

3. Environmental Impact Assessment (EIA) and Efforts to limit negative impacts of LNG on the environment

EIA that began to be used in the 1960s as part of a rational decision making process has become the key component of environmental planning and decision making is now being used increasingly in many countries around the world. At the onset, it involved a technical evaluation that would lead to objective decision making, and as practiced today as well, it is being used as a decision-aiding rather than a decision-making tool (Jay et al. 2007). In the specific case of LNG projects, EIAs have been greatly helpful in limiting environmental impacts and due to appropriate and modern equipment, facility design, safety and emergency systems, operational procedures and personnel training, the safety record of LNG operations could further be greatly enhanced, especially as there has been no off-site property injury or damage in 30 years[38]. The desire to limit negative impacts of LNG features prominently in the corporate statements of all LNG companies.

The sea transport of liquefied gases in bulk is internationally regulated - with regard to safety through standards established by the International Maritime Organization (IMO) and these standards are set out in the IMO's Gas Carrier Codes, which cover design, construction and other safety measures for ships carrying liquefied gases in bulk[39]. In recent years, in addition to internal innovations in their actions and equipment, LNG operators also benefit from innovative research by individual researchers and international organizations who have designed guidelines and methods for impact mitigation and benefit enhancement of LNG. Two important examples include: the General Environmental Health and Safety (EHS) Guidelines of the World Bank Group which recommends measures to prevent and respond to environmental and safety issues associated with LNG development (IFC-WBG 2007). There is also a framework for assessing hazards and identifying approaches to minimize the consequences to people and property from an LNG spill over water, presented in a report by Luketa et al. (2008), elaborating the general scale of possible hazards from a spills from 125,000 m^3 to 150000 m^3 class LNG carriers. This therefore means that in-depth knowledge and information currently exist on how to limit LNG impacts on the environment. Moreover, LNG companies are proposing and continuously developing new approaches to enhance the environmental acceptability of LNG operations. Specific cases are explained in the following paragraphs, based on a few localized examples, but which have been observed in numerous other LNG projects.

---

[38] Centre for Energy Economics - CEE, October 2003

[39] Retrieved online from http://maritimesun.com/news/wp-content/uploads/2012/02/GasTankersFamiliarisationCourse.pdf ; article titled Gas tankers: Familiarization course

New technologies have greatly reduced the number and size of areas disturbed by drilling, sometimes called "footprints." Satellites, global positioning systems, remote sensing devices, and 3-D and 4-D seismic technologies, make it possible to discover natural gas reserves while drilling fewer wells. Plus, the use of horizontal and directional drilling makes it possible for a single well to produce gas from much bigger areas than in the past, and the recent use of floating terminals[40].

LNG waste generation and management strategies: In the environmental Impact Statements of Santos LTD GLNG project, emphasis is made on the attempt by the company to tackle issues related to waste generation and management associated with the company, through recovering, re-using, recycling and efficient utilisation of all resources[41]as well as a Waste Management Plan (WMP) which provides a framework for waste management on site, through developing strategies for monitoring performance, setting performance targets and implementing action plans for improvement[42].

Ballast water impact mitigation: As a means of limiting the impacts of ballast water on coastal ecosystems, the Peru LNG project specified that ship tankers discharge ballast water 200km from the shore-line, significantly greater than the 5km stated in Peruvian regulation[43].

Coastline protection strategies: The "Gorgon Gas Development Revised Proposal", proposed the re-configuration of the Marine Offloading Facility (MOF) to reduce the need for drilling and blasting hard rock in the near-shore area off Town Point, as well as other measures to better protect the ecology and environment (Chevron Australia Pty Ltd, 2008)[44].

Rigorous evaluation of LNG proposals prior to implementation: In recent years, numerous proposals for the establishment of LNG terminals or other facilities are subject to rigorous evaluation. In the U.S., for example, the proposals of over 30 companies for new LNG terminals in North America, along the U.S. coastline or offshore were each rigorously evaluated before an LNG terminal could be constructed or expanded (USDOE, 2008). Likewise, the USCG and MARAD initiated the public scoping process on November 21, 2006, with the publication of a Notice of Intent (NOI) to prepare an EIS in the *Federal Register*. Thus, as a result of these developments, for more than 40 years, the safety record of the global LNG industry has been excellent, due to attention to detail in engineering, construction, operations (USDOE 2008; Foss 2007), as well as legislation and governance.

---

[40] Consumer Energy Report, Article titled Natural Gas Storage and Delivery, Effect on the Environment. Retrieved on http://www.consumerenergyreport.com/research/natural-gas-storage-and-delivery-effect-on-the-environment/ 8/21/2012

[41] Physical Effects—Water Intake And Discharge Facilities: Chapter Eight, retrieved from http://www.nefsc.noaa.gov/publications/tm/tm209/pdfs/ch8.pdf ;accessed in December 2011

[42] Section 5: GLNG PROJECT; retrieved online from: http://www.glng.com.au/library/EIS/05%20Waste%20Management%20%28Section%205%29%20FINAL%20PUBLIC.pdf ; accessed in January 2012

[43] ENVIRONMENTAL AND SOCIAL STRATEGY: Peru LNG Project http://www.perulng.com/ ; accessed in January 2012

[44] Gordon Gas Development Revised Proposal, Western Australia: Chevron Australia Document Control Number: G1-NT-FRMX0000062; Referral of Proposed Action vJAN08

It is also observed that some of the social effects such as accidents are human errors that could ensue from any other activity, no matter how benign. It should thus be understood that most of the issues enumerated by critics are usually taken into consideration during the EIA stage of LNG projects. Some LNG project proposals and/or later revisions usually spell out clearly where the major impacts may be felt, for example, Calypso LNG Deep-water Port Project of 2007 stated that impacts would be incurred by marine habitats (water column, benthic habitat, and areas of special concern), threatened and endangered species, non-threatened and non-endangered species, plankton, fisheries resources, and essential fish habitat (EFH), and proceeded to propose strategies for mitigation and/or elimination of deleterious impacts.

Papua New Guinea (PNG) presents a successful example in which extensive EIAs have impressively been done prior to the multi-billion Kina LNG project, incorporating in significantly equal proportions economic, social, cultural and ecological needs. The environmental impacts (EIs) that comprise extensive environmental, social and public health impact assessments and reviewed factors such as community needs, sensitive environmental habitats, and biodiversity is a more than 6 thousand page document that draws upon 26 supporting studies that took nearly two years to complete in 2009. The EIs was conducted to identify any potential impacts on the environment and the people of the project area and to enable the project to implement policies and procedures to mitigate adverse impacts and enhance beneficial ones. The social, cultural and economic objectives of the PNG LNG Project are to deliver a project that has the potential to benefit local and wider communities with minimal impact on the environment, by:

- Recognizing local amenity, values and culture of people directly affected by the project.
- Managing environmental and social impacts.
- Contributing to the state's economy and human wellbeing[45].

Despite the fact that many of impacts of LNG can be reduced or eliminated through the use of various techniques, procedures or technologies, as well as careful siting of the facility and timing, characteristic most LNG projects, social impacts on affected communities were only partially addressed.

4. Social Impacts of LNG Projects

There is growing dissent on the use of EIA as its influence on development decisions is limited as well as on the basis that it generally falls short of its full potential, because of the following reasons such as their narrow spatial and temporal scope (Lenzen et al. 2003) as well as focus

[45] http://www.pnglng.com/commitment/environment.htm accessed on 08/31/2012

generally on environmental and economic considerations (Yemen LNG Project EIA[46]). More importantly, EIAs have very narrowly focused on social impacts, due to the prevailing view is that money could compensate for any adverse social impacts (ICOGPSIA[47] 1994). Even when entire neighbourhoods have to be displaced, operators are generally misled by the belief that providing comparable or better housing elsewhere will solve any social concerns. In addition, social benefits are also generally narrowly defined in terms of their potential to increase employment opportunities and their ability to create a significant number of jobs during both construction and operation. For example, in the EIA of Peru LNG project, approximately 35,000 direct and indirect jobs created during construction and 2,750 jobs during the life of the project)[48], as well as boost national GDP.

This narrow econocentric approach to EIAs tends to ignore less tangible but extremely important aspects of community livelihoods, such as the spiritual and ancestral bonds to the land, attachment to nature and traditional ways of life, psychological and emotional impacts of land grab and developments.

Also lost in this process is the importance people attach to their communities and neighbourhoods, and particularly to long-standing social networks that form the basis of support both for daily living especially during periods of extreme stress and hardship. Moreover, during the construction of the breakwater and other components of the marine facilities, artisan fishing activities may be affected, as access will be limited and fish may leave the area due to construction disturbances, as well as envisioned social impacts associated with fishermen compensation and employment expectations (Reef n.d). The loss of significant revenue from the fisheries and tourism sectors, with the resultant loss of thousands of jobs, for example, a loss of more than $100 million a year commercial fishing industry in exchange for a mere $3.9 million in income from the 35 to 40 post-construction LNG jobs, was the assessment made of the Columbia River and Coos Bay[49].

In the U.S., numerous state agencies from Oregon Department of Environmental Quality (DEQ) raised serious concerns about Oregon LNG project impacts, and a diverse coalition of conservationists, farmers, commercial and sport fishermen, business leaders, private property rights advocates, and Native Americans are opposing one or more of the LNG projects[50], due to

---

[46] http://www.yemenlng.com/ws/en/go.aspx?c=soc_Environment

[47] Acronym for The Inter-organizational Committee on Guidelines and Principles for Social Impact Assessment, formed in the U.S by a group of social scientists, with the purpose of out-lining a set of guidelines and principles that will assist agencies and private interest in fulfilling their obligations under NEPA, related authorities and agency mandates. You refer to this originally on Page 19.

[48] Environmental and Social Strategy: Peru LNG Project http://www.perulng.com/ accessed on 05/20/2012

[49] (*See* article titled: Columbia River ports wary of LNG plans, Longview Daily News, December 13, 2008).

[50] Natural Gas Ports and Pipelines : Putting Our Economy , Communities and Quality of Life at Risk available online at www.Nolng.net

severe social impacts as well as serious impacts on commercial shipping caused by security buffers around LNG tankers, adversely affecting commercial and recreational fishing, tourism, farming and operations along the proposed pipelines[51].

In the Camisea gas fields in Peru, the LNG gas project has opened up one of the most pristine rainforest valleys in the Amazon, threatening the livelihoods of riverine indigenous communities and the physical survival of isolated indigenous populations[52].

Similarly, in Timor L'este, most of the people make their living from agriculture, using land not only to live on, but also for farming, as the source of their livelihood, and as part of their traditional spirituality. The Timor L'este LNG project plant was estimated to have a huge impact on community rights to land due to the displacement of numerous people, as well as the loss of their sources of livelihood, because their land is taken away to build the LNG plant and associated infrastructure. Moreover, the remaining farmland was assessed to be in danger of being polluted; sources of water diverted or contaminated by project activities, and the misuse and destruction of sacred places, infrastructure of communities living on or near the project site, thereby violating traditional beliefs, which are fundamental to Timorese culture (La'o Hamutuk's Natural Resources Team 2008).

The social limitations of traditional EIA therefore have the potential to spark-up severe social and even political tensions between communities and the LNG developers. Thus, in most LNG projects, developers, despite concluding that no major negative environmental impacts were likely to occur, there were however instances of strong opposition by some local communities. The major reason being that local socio-cultural, emotional and spiritual concerns are often completely left out of EIAs. Despite the completeness of PNG-LNG project EIA, criticisms arose, based not on social and/or ecological issues, but on the sharing of benefits between foreign investors and local communities. According to theses dissenting voices, based on the business deals, foreign investors will take away much of the Profits. In order to buttress their point and call the population into action, a website was designed titled "Act now! For a Better Papua New Guinea"[53], to raise awareness on this issue.

Therefore, by providing social infrastructure and services as part of their corporate social strategy, LNG developers often harbour the false impression that social issues are taken care of, and expect local communities to be satisfied, without actually consulting and involving them in

---

[51] Natural Gas Ports and Pipelines : Putting Our Economy , Communities and Quality of Life at Risk available online at www.Nolng.net

[52] Liquefied Natural Gas: The Wrong Choice for the West Coast. *Energy, 94104*(415)(2004)

[53] http://www.actnowpng.org/content/png-biggest-loser-under-lng-project accessed on 08/31/2012

decision making. Analysis of the general social impacts associated with, but not limited to LNG projects have been made in the following section, thus necessitation the call for stringent SIA to be incorporated into all stages of LNG project impact assessment, as well as other projects with similar far-reaching consequences.

This therefore necessitates the inclusion of two extremely important aspects in LNG projects: the stringent use of SIAs and effective stakeholder participation. In addition, SIAs should necessarily be conducted by SIA specialists contracted by the local communities using specific methods and tools, and not by the general EIA experts. This inability of EIAs to fully address social impacts reveals the incompleteness of project impact assessments, and thus masks the full picture of project impacts on stakeholder communities and other socio-professional stakeholders especially in the agricultural, fisheries and tourist sectors. There is even less concern for the distribution or "equity" of these impacts on different populations.

5. Final Remarks on the LNG debate

Effectively, three issues have been raised in this review. The first is the great rivalry existing between LNG projects and local communities. Second is the observation that LNG's are not always beneficial to affected communities, and third is the weakness in EIA's to effectively incorporate SIA principles. The following general observations have been made to summarize the main points of this paper:

- It is common knowledge that attitudes before implementation of LNG projects (as well as other large scale projects) often predict attitudes afterwards, as there is growing data detailing that fears are often overblown and that expected (often promised) benefits often fail to meet expectations. However, as mentioned earlier, concerns expressed by critics of LNG are not often sufficient to lead to a complete halt in LNG operations for two main reasons:
    i. In the current state of the global energy, especially with highly volatile oil prices and uncertainties associated with the future of oil; energy diversification is what is needed, and LNG appears to be doing just that.
    ii. Moreover, it has been demonstrated that ecological and environmental impacts of LNG can be taken care of with environmentally friendly technology and better adapted operational methods.
- Judging from the exponential growth of the LNG sector, as well as the support the fuel enjoys within the political elite, it can be confidently stated that LNG is here to stay, and criticisms won't do much to stop its growing importance and acceptance. Therefore, rather than pointing accusing fingers, emphasis should be on identifying problems

associated with LNG projects, and proposing appropriate solutions to deal with them, and in the process, dispel actual or potential threats.

- Though LNG developers have the means to deal with negative impacts of their operations, one category of impacts, of a socio-cultural nature, remain poorly understood, not only in the LNG sector, but also in other large scale projects, and therefore, not effectively accommodated in project impact statements. Therefore, there is need for a comprehensive impact assessment (pre- and post-) on a case by case basis, with SIA occupying an equally central spot as its environmental counterpart, in order to address all spheres of the misunderstandings and conflicts. To further stress the importance and urgent need for SIA, it is imperative to provide a brief account of the current practice of EIA and its shortfalls.

With regards to the rivalry between communities and LNG projects, an important observation made from most of the claims in opposition to LNG is that they are mostly based not on a balanced assessment of the pros and cons of LNG projects and recommended energy options, but generally on one sided perspectives driven by diverse motives which sometimes are purely emotional and conservative, and in other cases are driven by stances of 'deep / radical environmentalism. The bottom line is that all human activities in the natural environment have some deleterious outcomes which may be injurious to some organisms or segments of a population in some way; this is not exclusive to LNG. Based on these negative views on LNG the advantages of the fuel as enumerated by proponents have been generally dispelled as myths[54], and critics have continuously voiced their opposition to proposed or ongoing LNG projects, with mixed results, in some cases led by environmental advocacy groups such as Greenpeace.

On a positive note however, vocal critiques of LNG proposals have been highly successful in driving innovative changes in LNG operations as well as have forced operators to continuously aim at maintaining good reputations, to make policy adjustments, technological and operational improvements. Therefore, it is also the view expressed in this paper that critiques of LNG, if well founded, and scientifically and morally grounded, should not only be regarded as mere obstruction but most importantly as forces of change and innovation. Therefore, to dispel LNG development, despite their immense socio-economic potentials, on environmental, ecological and safety grounds is illogical because LNG developers have the means – financial and technological – to minimize the impacts of their actions, as elaborated in the following paragraphs.

[54] Liquefied Natural Gas : The Wrong Choice for the West Coast. *Energy, 94104*(415)(2004)

There is therefore a need for stronger foundation of EIA practice through training for practitioners, guidance on EIA practice and continuing research (Jay et al. 2007), in order to fill up these current gaps, as well as extend the scope of EIA to include indirect diffused impacts, and especially including social impacts as an integral part of the assessment. Moreover, all projects undertaken in highly sensitive social and ecological environments, which were implemented based on recommendations of traditional EIA should be reviewed and subjected to SIA as well, in order to identify areas in need of policy and project adjustments. Based on the results of such an assessment, mitigation measures may be taken by the project owners and/or the local governments in collaboration with impacted communities to limit community vulnerability and enhance social benefits and wellbeing, before and/or after a project has been implemented.

5.1. Principles of Social Impact Assessment (SIA)

As defined by the Inter-organizational Committee on Guidelines and Principles for Social Impact Assessment, "social impacts" refer to the consequences to human populations of any public or private actions that alter the ways in which people live, work, play, relate to one another, organize to meet their needs and generally cope as members of society.

The term also includes cultural impacts involving changes to the norms, values, and beliefs that guide and rationalize their cognition of themselves and their society[55]. In addition, it is important to consider the social equity or distribution of impacts across different populations. Just as the biological sections of EIS's devote particular attention to threatened or endangered plant and wildlife species, the socioeconomic sections of EIS's must devote particular attention to the impacts on vulnerable segments of the human population, for example, occupational, cultural, political, or value-based groups for whom a given community, region, or use of the biophysical environment is particularly important.

There are at least three benefits of identifying un-resolvable social impacts that may result from a proposed project as follows;

i. Identifying methods of compensating individuals and the community for unavoidable impacts
ii. The community may identify ways of enhancing other quality of life variables as compensation or the adverse effects.
iii. Identification of un-resolvable social impacts makes community leaders and project proponents more sensitive to the feelings of community residents (ICOGPSIA 1994).

[55] ICOGPSIA,1994

By articulating the impacts that will occur and making efforts to avoid or minimize the adverse consequences, or compensating impacted groups or the communities for losses suffered, benefits of LNG may be enhanced and avoidable conflicts can be managed or minimized. The fact that most large scale projects around the world do not give prime consideration to the social wellbeing and cultural values of the impacted communities due to ignorance or nonchalance, especially when such projects alter their living spaces and their livelihood activities, has also been exposed. In some cases, LNG company publications elaborate important social programs as part of their corporate social responsibilities, for example Oman LNG LLC Social Investment Programme (SIP) created in 2000 as an umbrella scheme under which it directs its corporate responsibility initiatives and financed by 1.5% of the company's post-tax profits. It is however important to find out whether the programs are implemented in close collaboration with the communities, and actually fulfil their socio-cultural wishes and livelihood aspirations. In a few cases where social impacts are considered, such considerations usually narrowly include only limited aspects such as employment, health, safety. In reality, it is easy for companies to execute social programs, just to fulfil their corporate social responsibilities before the watching eye of the international community, without such projects actually benefitting the grass-root populations, either because the impacted populations are not effectively included in the projects' design, and/or the social programs are not tailored to their needs. Therefore, in such a cases, only through a stringent social impact assessment can answers be obtained, and can such gaps filled.
More-so, even in countries which have prospered as a result of the development of energy resources, and displaying visible signs of economic affluence, SIA is still vital, as it may uncover malignant social or cultural issues afflicting a specific local community or socio-professional group in no small way (physically, psychologically, culturally, relational, occupational, spiritual), un-noticeable in the general illusion of abundance and satisfaction. This is in line with an important principle of SIA.
An important part of developing quality policy especially of an environmental nature, is ensuring adequate attention is given to providing accurate information to policy makers to assist their decision making (Jollands et al. 2003) To achieve this, making use of indicators is a popular decision support tool for environmental policy (United Nations 1995; OECD 1998). An indicator is a quantitative or qualitative measure derived from a series of observed facts that can reveal relative position in a given area and, when measured over time, can point out the direction of change (European Commission Joint Research Centre (n.d)). Recently, emphasis has shifted to combining indicators into groups or composite indicators to summarize a number of underlying individual indicators or variables, vital to assist decision-makers by reducing the clutter of too much information, thereby helping to communicate information succinctly and efficiently (Alfsen and Saebo 1993; Williams 1994; van den Bergh 1996; Callens and Tyteca 1999; Gustavson, Longeran et al. 1999); thereby, "necessary to keep from overwhelming the

system at the higher levels of the hierarchy" (Meadows 1998, p. 22). The use of carefully selected indicators is therefore central to effective successful SIA.

Given that the selection of indicators and the conduct of SIA are meant to serve as a decision support tools, Boisevert et al. (1998, pp. 106-7) summarise the nature of indicator information meant , to aid the comprehension of decision makers, as follows:

• Only a limited number of indicators should be used to convey the general state of the environment. Too many indicators can compromise the legibility of the information.

• Information should be presented in a format tailored to decision-making. This requires the construction of indicators that reduce the number of parameters needed to give precise account of a situation.

• Indicators should concentrate on the issue under consideration (Jolands et al. 2003).

Therefore, SIA for LNG projects must necessarily be guided by a number of carefully selected social indicators, categorized broadly under specific headings.

Moreover, the selection of indicators and actual conduct of SIA must be within the purview of well-established principles, as shown on Table 1, below.

Table 1. Principles of Social Impact Assessment

Principle	Description
Involve the diverse public	*Identify and involve all potentially affected groups and individuals*
Analyze impact equity	*Clearly identify who will win and who will lose and emphasize vulnerability of under-represented groups*
Focus the assessment	*Deal with issues and public concerns that really count, not those that are just easy to count*
Identify methods and assumptions and define significance	*Describe how the SIA is conducted, what assumptions are used and how significance is determined.*
Provide feedback on social impacts to project planners	*Identify problems that could be solved with changes to the proposed action or alternatives.*
Use SIA practitioners	*Trained social scientists employing social science methods will provide the best results.*
Establish monitoring and mitigation programs	*Manage uncertainty by monitoring and mitigating adverse impacts.*
Identify data sources	*Use published scientific literature, secondary data and primary data from the affected or potentially affected area.*
Plan for gaps in data	

Source: ICOGPSIA (1994)

It is the central view expressed here that specific focus of the Social Impact Assessor should be on the socio-cultural attributes not normally considered in traditional EIAs of LNG projects and often overlooked, not just the easily observed and understood ones, including but not restricted to spiritual and ancestral bonds, attachment to nature and traditional ways of life, psychological and emotional impacts, alongside livelihood activities, health concerns, etc, depending on the local realities of specific areas.

The Inter-organizational Committee on Guidelines and Principles for Social Impact Assessment alludes specifically to pre-project impacts, in terms of efforts to assess or estimate, in advance, the social consequences that are likely to follow from specific policy actions (including programs, and the adoption of new polices), and specific government actions. However, in the case where a large scale project has already been implemented without effective SIA, such as most LNG projects mentioned here, the guidelines can equally be valuable in assessing post-impacts, so as to guide policy and operational adjustments.

Therefore, to predict what the probable impact of a development will be, in pre-project social impact assessment, experts must try to understand the past behaviour of individuals and communities affected by agency actions, development, or policy changes (ICOGPSIA, 1994). Similarly, post-project social impact may be evaluated by comparing social conditions before and during/after the project implementation. In the case of proposed actions that involve controversy, attitudes and perceptions toward a proposed policy change are one of the variables that must be considered in determining the significance of impacts (ICOGPSIA, 1994).
It must be noted however that the principles and applications of SIA elaborated here are not unique to LNG. They can be conveniently applied and recommended for other projects which have the potential to cause significant changes to the biophysical and socio-cultural environments.

6. Conclusion

From the foregone, this paper has demonstrated that LNG development is a highly controversial issue, as observed in the views of both proponents and opponents, based on their specific interests. Analysis of the current debate on whether LNG is needed or not, is necessary so as to shed light on the concerns of both parties involved, thereby narrowing the rift and encourage harmonious coexistence with LNG developments. In order to promote such coexistence, it is therefore the responsibility of LNG companies to overhaul their current practice of corporate social responsibility through the adoption of stringent SIA at all stages of their operations, on a spatio-temporal basis. The findings of SIA have the potential to prompt policy adjustments and technological and operational innovations in order to ensure a broader, mutually beneficial relationship between the LNG sector and other sectors of human existence as well as ensure the sustainable development/livelihoods of communities. Therefore, based on the Environmental, Health, and Safety Guidelines developed by the World Bank Group, it can be cautiously stated that if LNG developers observe the guidelines closely, apply state of the art technology and strategies, an effective governance framework that is integrated and interactive, involving multiple stakeholders, as well as an extensive assessment of social and livelihood impacts on potentially affected communities, the adverse impacts of LNG as cited by opponents can be greatly minimized and/or eliminated. Moreover, in the context of such conflicting views on LNG, it has been generally noted that attitudes (particularly negative ones) formed about a project cannot be eliminated, but could be moderated if the public has complete information about the proposed development, are included in the decision-making process, or provided with structural arrangements that assure safe operations. In such a case, LNG development will assume its original intent as a fuel for the future, contributing to social-economic, environmental and energy sustainability.

References

Al Busaidy, E. (2008). Liquid Asset: the rise and rise of LNG. Al manhal number two, Mazoon Printing, Publishing & Advertising (L.L.C)

Alfsen, K. H. and Saebo H. V. (1993) "Environmental quality indicators: background, principles and examples from Norway."Environmental and Resource Economics 3: 415-435

Al Ghanjah (2011)– march: A newsletter produced quarterly by Oman LNG L.L.C. under the General Supervision of Nasser Al Kindi, Communications, Issues and Crisis Management Department, Oman LNG, P.O. Box 560, Muscat 116, Sultanate of Oman; Issue1 of January

Boisevert, V., Holec, N. et al. (1998). Economic and Environmental Information for Sustainability. Valuation for sustainable development: methods and policy indicators. S. Faucheux and M. O'Connor. Cheltenham, Edward Elgar Publishing Ltd: 99-119.

Branch, K., Hooper, D.A., Thompson, J. and James C. C. (1983). Guide to Social Impact Assessment. Boulder, CO: Westview Press, ISBN 0-86531-717-8.
Brudge, R.J. (1994). A community Guide to Social Impact Assessment, Middleton, WI (P.O. Box 620863): Social Ecology Press, 210 pages.

Callens,I., Tyteca,D.(1999)."Towards indicators of sustainable development for firms." Ecological Economics 28(1): 41-53

Cedigaz (2006) BP Statistical Review of World Energy

Cedigaz (1970 - 1992) Statistical Review of World Energy

Environmental Impact Statement of PNG LNG Project (n.d). Chapter 20: Environmental impacts and mitigation measures: LNG facilities. Coffey Natural Systems 1284_9_Ch20_v3.doc 20-1, PGGP-EN-SRENV-000001-001 Rev0

INPEX Corporation Ichthys LNG Project Final EIS - 2011. Community concerns addressed in final Environmental Impact Statement for Ichthys LNG Project. Public Relations Group, Corporate Communications Unit Akasaka Biz Tower, 5-3-1 Akasaka, 19 April 2011

European Commission Joint Research Centre (n.d). Composite Indicators: An information server on composite indicators and ranking systems. Econometrics and Applied Statistics Unit. Retrieved online at: http://composite-indicators.jrc.ec.europa.eu/S10_visualisation.htm

[FERC] Federal Energy Regulatory Commission (2005). Draft environmental impact statement, Down Landing LNG and Logan Lateral projects. Washington (DC): FERC Office of Energy Projects

Foss, M.M. (2007). Introduction to LNG: An overview on liquefied natural gas (LNG), its properties, organization of the LNG industry and safety considerations. Updated January 2007, Centre for Energy Economics - CEE

Gorgon Gas Development Revised Proposal, Western Australia (2008): Chevron Australia Document Control Referral of Proposed Action Number: G1-NT-FRMX0000062

Chevron Australia Pty Ltd, 2008. Gorgon Gas Development Revised and Expanded Proposal: Barrow Island Nature Reserve. Environmental Protection Authority Perth, Western Australia; Report and recommendations of the Environmental Protection Authority, Report 1323

Gustavson, K., Longeran, S. (1999). "Selection and modelling of sustainable development indicators: a case study of the Fraser River Basin, British Columbia." Ecological Economics 28(1): 117-132

[IFC-WBG]The International Finance Corporation and the World Bank Group (2007) Environmental, Health, and Safety Guidelines for Liquefied Natural Gas (LNG) Facilities. *Hazardous Materials Management*, 1-19.

[ICOGPSIA] The Inter-organizational Committee on Guidelines and Principles for Social Impact Assessment (1994). Guidelines and Principles for Social Impact Assessment; U.S. Department of Commerce, National Oceanic and Atmospheric Administration; National Marine Fisheries Service

India Dahej Liquefied Natural Gas Terminal Expansion Project, July 2006, Project Number: 39921
Environmental Impact Statement Prepared by Petronet LNG Limited for the Asian Development Bank (ADB)
International Gas Union (IGU) (2011) World LNG Report 2011. News, views and knowledge on gas – worldwide. Sponsored by PETRONAS. KL Malaysia

Jay, S., Jones, C., Slinn, P., Wood, C (2007). Environmental Impact Assessment: Retrospect and Prospect Environmental Impact Assessment Review 27: 287-300

Jollands N., Lermit J., Patterson, M (2003). The usefulness of aggregate indicators in policy making and evaluation: a discussion with application to eco-efficiency indicators in New Zealand. New Zealand

Kanagawa, T (2007). Japan's LNG Utilization and Environmental Efforts; The Japan Gas Association; retrieved online at: http://www.gas.or.jp/english/letter/images/07/pdf/LNG_Workshop.pdf

La'o Hamutuk's Natural Resources Team (2008). Sunrise LNG in Timor-Leste: Dreams, Realities and Challenges. Timor-Leste Institute for Reconstruction Monitoring and Analysis

Lenzen, M., Murray, S., Korte, B., Dey, C (2003) Environmental impact assessment including indirect effects – a case study using input-output analysis, Environmental Impact Assessment Review, 23, pp. 263-282

Luketa, A., Hightower, M., & Attaway, S. (2008) Breach and Safety Analysis of Spills over water from Large Liquefied Natural Gas Carriers; Retrieved from http://lngfacts.org/resources/SANDIA_2008_Report_-_Large_LNG_Vessel_Sa.pdf

Meadows, D. (1998). Indicators and information for sustainable development. USA, The Sustainability Institute

Mosolova T., Dyomkin T. (2009). "Russian LNG plant extends Kremlin's energy reach". Reuters. Retrieved 2009-02-19

Organization for Economic Co-operation and Development (1998). Towards sustainable development - environmental indicators. Paris, OECD

Oman LNG L.L.C (2010 (1)). 'Caring for the Environment: Partner in leading national growth'. Bak Publishing

Oman LNG LLC (2010 (3)). Social Investment Program; Partner in leading national growth

Oman LNG LLC (2010 (4)). Fact-sheet on Oman LNG; Partner in leading national growth

Oman LNG L.L.C. (2010 (2)). Empowering people, Building the future: the Oman LNG Social Investment Program, Bak Publishing

Reef, B. (n.d.). 10 Reasons why the Browse LNG development should not go on the Kimberley coast: www.wilderness.org.au/kimberley accessed in December 2011

Sakmar, S.L. (2010) The Globalization and Environmental Sustainability of LNG: Is LNG a Fuel for the 21st Century?; World Energy Congress, Montreal, Canada, September 12-16,

Taylor, C. Nicholas, C. Hobson B. and Colin C.G. (1990) Social Assessment: Theory, Process and Techniques. Studies in Resource Management No. 7, Centre for Resource Management, Lincoln University, New Zealand.

United Nations (1995). Work programme on indicators of sustainable development of the Commission on Sustainable Development. New York, United Nations.

[USDOE] U.S. Department of Energy (2008), in collaboration with the National Association of Regulatory Utility Commissioners (NARUC); Report titled: Liquefied Natural Gas: Understanding the basics

[USDOE] U.S. Department of Energy (2004). FE-LNG Imports by Country of Origin

[USEIA] U.S Energy Information Administration (2005). *Annual Energy Outlook*

Van den Bergh, J. (1996) Ecological economics and sustainable development. Cheltenham, Edward Elgar Publishing Limited.

Williams, M. R. (1994). Use of principal component bi-plots to detect environmental impact. Statistics in Ecology and Environmental Monitoring, University of Otago, University of Otago Press